# Oral Mathematics A Step in the Right direction

**We learn only to be of help to ones who need us.**

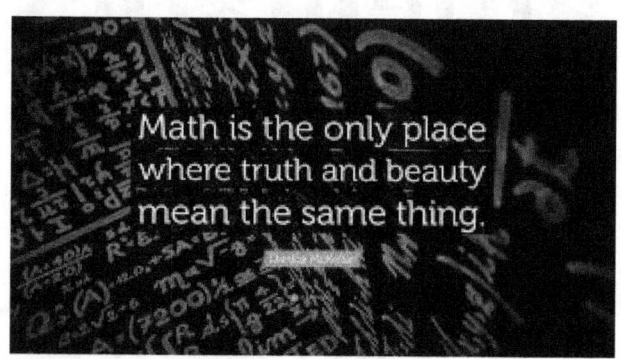

1. *Mathematics is not about numbers, equations, computations, or algorithms: it is*
    *about understanding. –William Paul Thurston*

2. *Life is a math equation. In order to gain the most, you have to know how to convert*
    *negatives into positives.*            *– Anonymous*

3. *Mathematics is the most beautiful and most powerful creation of the human*
    *Spirit.*            *--- Stefan Banach*

4. *Mathematics is, in its way, the poetry of logical ideas. –Albert Einstein*

5. *Mathematics knows no races or geographic boundaries; for mathematics, the*
   *Cultural world is one country.*          *-David Hilbert*

6. "Since the mathematicians have invaded the theory of relativity I do not
    understand it myself any more."                    —
Albert Einstein

7. "But in my opinion, all things in nature occur mathematically."
                                              — Rene Decartes

8. "With me, everything turns into mathematics."
                                              — Rene
Descartes

9. What is mathematics? It is only a systematic effort of solving puzzles posed by         nature.
------Shakuntala Devi

10. Mathematics allows for no hypocrisy and no vagueness.
                                              — Marie-Henri
Beyle

Stuck on numbers; they wish they knew their marbles. We were in Kenya; suffering indecision we could resolve if only we knew our numbers and handy solutions.

## PREFACE

Do we remember when first we learned to count to ten or when first we learned the alphabets it looked almost like an uphill task. We thought we had arrived and were at par now with all the great philosophers of our times; it was a thrill of our lives; we thought we can count away our troubles now one by one; no one can outwit us anymore. Wrong! It does not work that way. At every turn of the century there are new questions, new issues that take the center stage and attack our mental landscapes all anew ushering in challenges unheard of before. It is like this Corona Virus, the so called COVID-19, it had its family here before; it literally killed folks by the hour; it has a different strain now, more lethal and more of a killer, it has taken the World by storm. They say Pestilence and pandemics are synonyms meaning this pestilence is all well explained in the Bible and it has arrived as stated or so it seems, at the time it was meant to arrive. It is no news to Bible scholars; they know it all. For those of us who are stranded here, conflict, concern and confusion constantly afflicts us but we will see the light at the end of the tunnel God willing soon. We struggle staring at the bill sitting in a bar or a restaurant with folks who are mere bar flies than anything else as they jump from table to table, corner to corner to enter into loose and livid discussions or random dialogues with anyone they come across or who would listen, with total strangers, half drunk paying the same bill over and over again; it is scary to see them in such lethargy and aimless wandering. **If only we could get out of this mess**, if only we could get it out of the way, **if only we knew some basic math** we could be back home watching our favorite game on our Big Screen TV instead of languishing in these smoke filled chambers we went in only for a little break and are still there after all these hours, we so badly wish we could be out. I watch folks struggling to divide simple numbers when a little light in their brains could do it all for them. Life is complex as it is and we are not that smart in the first place when we bank on Artificial intelligence rather than on our own skills; we feel good we are blessed nonetheless. We need not be afraid of making a mistake.

Mathematics is not just a group of numbers stacked together or a book of handful of equations of lengthy iterations; it runs into every nook and corner of our Lives; it plays and directs every area of our lives. From music to operating theaters in surgical wards of hospitals, from waterways to dry land we live on; from the limping slow pedestrians to Banking of Roads, from Earth to the empty spaces in the atmospheres above, it plays its parental pivotal role in helping focus on matters at hand and helps chart our courses appropriately to help us land and reach our destinations in one piece. If you could go back in time when we first landed on Moon back in 1969, late Neil Armstrong said it so well that for the first time he saw everything so well knitted and in order; how everything was so well timed and fell in place; how exact was rotation of moon around the Earth and all that and his famous quote "That's one small step for man, one giant leap for the mankind."

It was all made possible by mathematics both Oral and documented. In Board Room meetings where time is at a premium we need ready answers and a little storage in our brains is called for to keep the ball rolling..

"That's one small step for *a* man...". that is what I said for sure, said Neil later.

Sir Isaac Newton would have been the happiest man were he alive today seeing us land on the moon seeing his laws of gravity and his numbers play a key role in this great shining success. Those curious how it all adds and holds up, how mathematics plays such a fatherly role, one just has to visit the sacred halls of congress here in the capitol where science flexes its muscle strengthened by principles of mathematics in its huge domes for all to see. You can hear your shouts from one end of the dome to the other at the exact spot on the other end; a good hundred yards separate the two points in the ceiling. People may not like to hear you ever but you can take solace in the fact your own voice does not go waste. It is only an echo much like the echoes between peaks in the remote hills and their deep valleys. The two points are nothing more than mere foci of an ellipse, not eclipse; we learned it all way back in our elementary classes if we paid attention then. Mathematics is the key to a healthy vibrant Life without it we will be sailing in murky waters.

**Math** and **music** are indeed related and we commonly use numbers and **math** to describe and teach **music**. These are all **mathematical**

divisions of time. Fractions are used in **music** to indicate lengths of notes. We should not have to be looking into our calculators all the time to feel the weight of those little fractions to know their sweep.

Churches regardless of their denominations design their sacred halls of worship to not only dictates of the Bible but also to the established facts of geometry to help music waft through the aisles like cool soothing clear water flowing in healing heavenly channels; logical catenaries define the layouts in accordance with accepted principles. In Surgical operations, Doctors use Laser Beams to melt kidney or bladder stone as they position their beams exactly at two points of an Ellipse (Focal Points); there cannot be anything safer than that. An Engineer would only be a fish out of water without its principles; he would be long dead without those answers that are needed in a flash. What am I saying? We need to know basics of mathematic. There is no sphere of Life that is run by anything other than principles of mathematics; we better learn them by heart in oral phases in daily exercises. Calculators are good but nothing can replace human intelligence.

We fall a prey to this Evil world almost daily; we do not know how to count or how to add orally and suffer for no reason. How easy it is to deceive one who is ignorant. Asian Landlords have robbed poor peasants and laborers all through the centuries and paid them little or nothing knowing they were defenseless and ignorant on top of it to know any better; tide has turned today and world is moving fast towards a new era where no one no more is ignorant. Greatest curse is in being ignorant and stay ignorant; GOD sent us down here fully armed with HIS tools. You do not need a calculator to know you are not being cheated.

The beauty of mathematics is its complete control of human logic and dimensions; it helps us to see how even a crooked path can be made straight; how a wayward system can be brought back to life after a little analysis; it all is based on its principles. Numbers play a great role in everything. When you say you are cold, it means nothing; when you say it is 20 degrees below normal it means everything. We have to have sense of knowing what these numbers mean and what role they play in our Lives. In this age of different units, if one says it is 30 degrees C out there, we should be able to see what it means in our own units. We do not have to multiply it by 1.8 and add thirty two to keep struggling, just multiply by two and add thirty two to get a round figure and see if you can be OK in that Heat. No need of a calculator!!!!! It is close enough.

It is not a one way street either. Those who deceive receive the same dose of medicine they administer to poor innocent souls, note souls, not folks for folks can react but souls are gone to have any more part here. In the absence of a basic understanding, even the most alert of salesmen lose half of their profits as they try to fleece their customers only to find later they had lost all they thought they had made as they sit to count their profits at the end of the day. If only they knew a little oral mathematics and could play it away in their heads before diving into their deals they could be in the driver's seat instead of crying over their spilt milk.

From Issac Newton, Archimedes, Carl Gauss, Leonhard Euler, Euclid, Lagrange, Leibniz to Galileo Galilei to even Omar Khayyam, the chain of logic has never ever been broken and what we have today is their ever clear mind telling us two and two will always add to four no matter what else may go wrong around the block.

In the language of mathematics 0+0 is always 0 meaning if you do nothing, you gain nothing and there are some hundreds of books on the subject; I saw one at a home of one of my professors, to a layman it may mean nothing yet it is the very base of all understanding in the world of logic. Arabs who are said to have invented this number (0) imagined it long before anyone even dreamed about it; sitting near their shallow ponds; they throw little pebbles out there and watch ripples recede in circles. Concept of a true zero; its true origin is too complex to describe here and can be looked into in any book.

Boolean algebra that deals in these 0s(zeroes) and 1s(ones) takes its name from its inventor Bool who invented this branch of mathematics for processing true or false values instead of just numbers. If your inputs are only 0 and 0 in the logic gates, you would only get a zero and hence we always are on the right track knowing we reaped what we sowed. Mathematics is way beyond counting numbers, it takes us everywhere in every field. Those of us who think parallel lines never meet, they do and they meet at infinity which for all intent and purposes is saying they don't meet in the sense we are used to but for ever alert scientists and old engineers it has its own significance. I used to be laughed off when I would make this unknown statement.

Greeks gave us the basic fundamentals of mathematics. We would not be here but for them. It is said Pythagoras was always busy drawing lines in the sand and when the invading armies came rushing with their horses almost close to erasing his work, he begged them to kill him rather than wipe out his lines he was working on. One other little story takes us to his 'Will' and determination where he refused to trample on the bean crop to escape from the hands of his killers and thus met his fate and died soon after. Though Babylonians had invented the right angle triangle before, he proved it later in his famous Pythagoras theorem and laid the foundations for many of its applications today.

What we are trying to say here is mathematics is more than numbers and has its role in every area of our lives. Oral mathematics as it is called used to be a main line subject in our courses one time where kids were made to memorize a few numbers before they could be promoted to the next levels. We are not stressing memorizing here; we are going the next step; setting the stage for true practical hands on computations.

I will be wanting in all of this if I left out a very important issue in our system of Education today. We have to be humans always no matter how big we go, read all stars and their galaxies, may know all orbits, and cover every mile around the globe. Here is a little note from a survivor of Holocaust that says Education in of itself is no ground for clean life. A demon can read and write, add and subtract and even a reach far beyond many of us but what good is that. He is out to devour, kill and destroy.

He says," I am a survivor of a concentration camp. My eyes saw what no person should witness. Gas chambers built by learned engineers. Children poisoned by educated Physicians. Infants killed by trained nurses. Women and babies shot by high school and college graduates. So I am suspicious of education. Reading and writing and spelling and history and arithmetic are only important if they serve to make us human". I am not a Holocaust survivor but I can take it to its next level. I had to face no Hitler in my life but ones I faced were no less evil; nothing may have changed since then. Folks today are no different and are perhaps more evil than the ones he describes above. They may not have to design Ovens to burn us alive but they do keep us on slow heat; they may not poison us but they do not let us eat in peace and on and on. Is it any less of a crime when you count someone out of his legitimate rights or pay him less than his due; is it less of a crime when you pay to get a degree or a Diploma from a school and head some Department of human Sciences or organized behavior and feel little or no compunction in dealing low blows and tortures to weaker people. It is the same. It is the same air, only its direction has changed. We must learn to quantify everything and Mathematics comes very handy in going there. If only you could add the number of crimes you commit daily or weekly or even yearly and can multiply the years you have been doing them using simple checks explained here, you would sit to think you have been no saint and may need to turn the corner. All I am saying here is do not keep things in a general mode, quantify everything in simple iterations, simple checks of your calculations would suffice  you may say out loud then, "Did I do that".

This whole effort is purely my own brainchild; I hear there are other similar attempts some other places but I have not seen them. If there are, it is a great gift to society for basic numbers should be well understood by all means necessary by every member of society. You can multiply any numbers with its help and check out your answers. I have chosen to only square them away. I have given two examples of multiplication in 72x92 & 53x83 for reference.

Try as many as you can and feel good once you have the handle on them. We are children of GOD playing on the sands of time seeking HIS knowledge  HE put us here below on earth for a reason; HE  is not surprised when we make a mistake, HE does not want us to repeat it.

**In mathematics all leaks can be captured. A wrong answer stares you in the face much like no water in the faucet tells there is a leak somewhere on way; absurd answers are so obvious you cannot miss them. Check them out before it is too late.**

Our purpose here in this little Book is to see if we can play with the numbers, multiply, divide, subtract or add them in daily use without shedding tears, without using little calculators. There is a book **Quality without tears** that talks of achieving Quality in any field without going mad. There were no calculators in Newton's time yet never such absurd answers were out there for gods to wonder if indeed these were humans they sent down here who were giving these dumb answers.. No one ever shouted Two Plus made five and not four. Calculators or Computers were never meant to supplant intelligence as is today the case. AI- Artificial Intelligence is doing much harm as its followers are beginning to find out that all is not well at its terminals. We quote all kind of successes everywhere with AI; it is shown defeating number one chess players of the world may be an exception than the norm.

This whole exercise is meant more for the day to day roadside traders than for the grocery shopping housewives; they have ample time to settle their bargains without losing their purses or currency they carry from all four continents; other folks too will find it equally useful though I doubt it for we now are so dependent on calculators or other Intelligence. You have to love mathematics to like this book.

We only tap into what GOD has already discovered. We only discover; hence credit goes to HIM and not us. He laid the foundations of the earth, HE separated waters from the Land and it is only HE who fixes all Times in our Lives. A time to be born and a Time to die, a time to rend and a time to sow is a great little piece in the book of Ecclesiastes saying strike the iron when it is hot; learn while you can.

I have taken biographical details of great mathematicians from the internet and their few quotes besides a fact from here or there, but the main body of the book is mine in all its details.

## Foreword

I taught mathematics all my life but never ever came across such clear approach to basics of mathematics. In the field of mathematics, checking your answers is so crucial that it can mean a change that can turn the tide as one wrong answer can throw you off way out completely; it can take forever to undo the damage ..

In a question in a test for licensure that asked amount of water needed to bring temperatures down from 50 Deg C to 10 Deg C as a result of dumping 100,000 lbs of Steam into a lake from condensers of a coal Power Station where fish swim and thrive; a kid came up with an answer so absurd he smacked himself hard on his head with his pencil seeing it in that low range. His teacher asked him to look it over again; he was surprised at his answer, he saw his folly. We know when we are wrong and at times hit the wrong buttons, what can we expect then. Oral mathematics and checks do come handy at times like they could be here.

I am very old now and so is my student and author of this work; I at 88 years and he at 78 years of age have no other desire except to see our young ones swift and alert in this field of mathematics and achieve their goals without much stress. We are behind most other nations in mathematics and science; we need to bolster our efforts to undo this lack and this little work is a step in that direction. 'Trust but verify' is a great slogan; President Ronald Reagan would often expound on it as he would deal with rogue nations; same line of thinking should be our way of life as we deal with complex queries and give our final answers. Happy computations! God bless us all.


Abigail Mason, Ph. D




## INDEX


1.  Squaring of Numbers ending in digit 1  ( last digit  1) 10-14


2   Squaring /multiplying-----------------do- 2--- 15-17


3.  Squaring------------------do-------------- 3 17-18


4.  Squaring------------------do------------- 4 18

5. Squaring-------------------do-------------5
18-19

6. Squaring--------------------------------6
19-20

7  Squaring--------------------------------7
20

8. Squaring--------------------------------8
-21

9  Squaring--------------------------------9
21-22

10. Squaring------------------------------ -10
22

i   Napoleon's theorem
23

ii  Morley's theorem
24

iii General
25-28

iii Six Great mathematicians
29-33

iv  Pascal
35

v   Roman Numerals
36

vi   Prime Numbers
38

vii  Conclusion
39

viii Additions / Ramanujan
40

## Chapter 1

### Squaring of Numbers ending in 1.

I. **Numbers ending in 1**: for example: Numbers 11, 21, 141, 1001 and so on and so forth. All have number 1 at their end.

Say we want to multiply 11 by 21

**11x21=? Remember reference number under review is 1.**

Let the last digit 1 stay as it is; so we insert 1 at the end in the answer moving from right to left. So the number 1 stays on the right just as it was.
First step: Number 1 stays as **1** (our reference base) --------
**C=1**
Next step: add the number on the left of 1, so 1+2=**3**-------
**B=3**
Multiply again the numbers on the left of 1, so 1x2=**2**-------
**A=2**
Hence the answer is 11x21=**2(A)3(B)1(C)=231**

Now let us check if it makes sense:
11------1+1= 2------------ i

21 ------2+1=3------------- ii

Now multiply i by ii: 2 times 3; answer is **6**. LHS=6
Check if our answer for 11x21=231 is correct. How?
Simple: add all digits on Right hand side: 2+3+1=6 : RHS=6
**LHS=RHS meaning 6 is equal to 6. OK.**

Let us try another big number like 121x141:
Now **12** in 121 and **14** in 141 are to be treated as units; We cannot separate **12** into 1 and 2 as also cannot separate **14** into 1 and 4.
First step: Number 1 stays put as **1** in final answer --------------------
---------------------- i
Next Step: Add numbers **12** and **14**=26: **Carry over is '2' for the next step,** retain **6 for final answer**-----------------------------------
---------------------------------------ii
Multiply the unit **12** by unit **14**, so answer is 168: plus the carry over **2** from step above:
so it is **168+2= 170** ------------------------------------------------
----------------------iii
Answer putting them all together is (iii-ii-i) from above= 170 6 1
Check if it is correct. Left Hand Side :
Add all numbers of **121**: so they add up to 2+1+1=4
Add all numbers of **141**: so they add up to 1+4+1=6
Multiply 4 times 6= 24: add the result: 2+4=**6**
**LHS=6**
**Now RHS: The Result:**
Our answer is 17061
Check the Answer:
Add all digits one by one: treat number 9 or its multiples like 9x2=18 as zero.
1+7+0+6+1= 15: any multiple of 9 or 9 itself, subtract it from the answer.
So number 15 is reduced to **6** after deducting 9 from it which is the same as 6 from above.

Its more use is found in the analysis of Electrical Circuits where we are called upon to square Resistances and Reactance to get to Impedance. $Z = (R^2 + X^2)^{1/2}$

Even Simple Pythagoras theorem which says Sum of the squares of the two sides of a right angle triangle is equal to the square of the hypotenuse; if the two sides of this right angle triangle are 3 and 4, the hypotenuse would equal 5 after applying the theorem.

$3^2 + 4^2 = 25$: so square root of 25=5. If the sides are 21 and 11, it sure will be a little tough to figure it out without a calculator in good time but it can be broken down.

But in our simple way $11^2$ is equal to 121 applying the same formula for numbers ending in 1.

And $21^2$ is equal to 441 by the same token.

So 441+121=562

Would you like to check if 11x11 is indeed 121. Let us try one more time.
First step: leave number **1** just as it is in the answer---------------------------- i
Next Step: Add the numbers to the left of 1 from both elevens =**2**--------------ii
Finally multiply numbers to the left of 1; so **1x1=1**----------------------------iii
Answer is=**121**

**And then final check:**

**11= 1+1=2**
**11= 1+1=2**
**2x2=4 --------------A**
**121=1+2+1=4----B**
**A=B: Hence our answer is correct.**

Note: Anytime a number becomes more than 9, subtract this nine or its multiples; like when it becomes 18(**9**x2) or 27(**9**x3). Both 18 or 27 are dumped or ignored then.

Going back to our earlier problem: 121x141; we had as an answer 17061; sum of all its digits exceeded our base reference number 9, hence we deducted 9 from 15 and answer was 6 for this iteration and from the problem itself 121x141, the outcome was the same as 6 and so our multiplication was correct. How?

121=1+2+1=4
141=1+4+1=6
4x6=24; 2+4=**6** (we could deduct 9x2=18, multiples of 9 to come to the same outcome).
Any numbers running into thousands can be split up into one unit while the number we base our procedure on remains the solitary number to the right of this unit.
Example:

Problem: **1051x1051**

**Our second unit now is 105 to left of our reference base number 1 Remember reference base is always 1 if we are trying to solve problems ending in 1.**
**It is always 2, if we are trying to solve problems ending in 2 and so on and so forth.**

So we would proceed in the same way:

1x1=**1**
105+105=21**0** (carry over is 21 here)
105x105 can then again be treated as units with number ending in 5 and easily multiplied on those lines.
105x105=11025-----( we can again treat it as number ending in 5 and use procedure used for the number 5 we will see later)
11025+21=**11046**
So the answer is **1104601**
Check it:
First Step: **1+0+5+1=7**---------------------------------------------i
Next Step: **1+0+5+1=7**---------------------------------------------ii
Next Step: 7X7=49--------------------------------------------------iii
Next Step: 4+9=13-9= 4-(any number above 9; subtract 9 from it)--A

Our Answer: **1104601**= 1+1+0+4+6+0+1=13-9=4----------------------
B
Hence LHS=RHS

Sample questions: **51x51: solve and check your answer as described above.**
          **181x181: solve as described above.**
          **171x161: solve as described above**
          **1051x1071: solve as described above.**

Check answers equating the data in the problem and seeing it equal to the answer arrived.
Data in the problem is the actual numbers we are trying to multiply. Answer is the result of multiplication, using method shown above check if the two agree.

Learning begins when you are young and eager. These little kids in a school in Kenya are on their way to learn their first few numbers. I saw them in Kenya, full of life, full of innocence and last but not the least full of discipline.

They know how to count now. They know how to add or subtract now. They are young and smart.

How wonderful it would be if they could master the art of these checks and balances shown in the example above at this young early age and do away with memorizing and storing numbers in their little memories.

Residents in Hospitals have stopped memorizing any data whatever they need to store in their tiny little brains or must when they are looking at the medical reports of their patients in all kinds of wards; they just scan their little hand held computers and know what is right and what is wrong; why then little kids are made to memorize numbers they can safely check with proven techniques of mathematics. Memory is a great gift; it certainly has its place but along with it we also need to be able to check our numbers using basic means. It gives us a sense of completeness.

**A school in Kenya; Little ones entering their classes**

## Squaring of Numbers ending in '2'

### II. Numbers ending in the digit 2

We are going to raise the bar and step into our next attack. We are going to go to our next integer and that is the number 2. We will try to evaluate all such iterations as involve numbers that end in this digit 2.

### Reference base Number 2:

For example; 1**2** x 1**2** ; 2**2**x3**2** ; 15**2**x17**2** etc.
Let us see: what does 12x12 come to?
12--take 2 from its right side; so we have 2------A
x
12--take 2 from its right side; so we have 2------ B

-----
144

First step: multiply 2x2 and leave it for use later **4**--------------C
Next Step: Add the integers **1+1**= 2 and multiply by the reference number 2:

$$2x (1+1) = 4 \text{----------------------------------------D}$$

Third Step : Multiply the same ones on left of 2:1x1=**1**------- E

Answer is 144. It is much easier to compute it this way. Numbers can go big at times

We can check it by the same means we deployed before to check our answers.

**Again what is our procedure; Elementary Dr. Watson as Sherlock Homes would say.**

**Multiply the reference base number 2 by the number in the second line below it**

**Sum the digits on the left of 2 in both the lines and multiply by reference base number:**

**Multiply the digits on left of 2 in both the lines; we then have our answer.**

Let us attempt numbers that are not squares of each other.

**Multiplication of numbers ending in '2'**

<u>72x92.</u>

**First step: 2 times 2 is 4.**
**Second step: 7+9 times 2 is=32 (carry over is 3)**
**Third step: 7x9 plus the carry over 3=66**
**Answer is 6624.**

**Let us check:**
**72=7+2=9**
92=9+2=11
9x11=99=9+9=18= multiple of 9, so ignore it is zero.
Our Answer:
6624=6+6+2+4=18=multiple of 9, hence Zero.
So our multiplication is correct.

We can go to any number. Apply the same procedure for numbers ending in the base reference number, in this case it is 2 and you can never go wrong. In engineering analysis, we need to be able to arrive at quick conclusions.

Key is to always keep the base reference number separate, the one used for computational access. All numbers start from 0 and end in 9 and hence help us devise easy means to evaluate them in their true habitat. If we are dealing with 21, our base reference number is 1, number 2 to the left of it is inserted as shown in the exercises and procedures well detailed above. If we are dealing with 145, our base reference number is **5**; procedures for numbers ending in 5 only apply and the number **14** to the left of it is treated as a unit rather than 1 and 4 separately to arrive at the answer.

If the multiplication is between 11x11; the lone number **1** on the extreme right is the base reference number and the number 1 on the left is the one to be worked on as usual and shown before.

Numbers are everything; here frequency on the meter is way below 50 C/S meaning action is needed; lines are overloaded and station breakers can trip. We need to act quickly to calculate the load with all its variations without the help of Calculators. Exercises in this little book may come handy to check our answers.

## III------------------Numbers ending in Digit 3 (Three)

Let us move on. We are going now to that ever green 3 (three). Three is an odd number, odd in every way, it is called the third wheel, the third nail; only thing in its favor is its reference to third base in Baseball where you can be assured you are nearing the home run having reached there; you may only need just one good hit to score your home run.

Let us try some numbers now.

Procedures for multiplying numbers ending in 2, 3 and 4 are exactly the same but we would take them on just to prove our point.

Let us take 23x23=529 or any number ending in 3.

How?

First step: Multiply the reference number 3 by other number in the equation, i.e. 3
        3x3=**9** ----------------------------------------------------------i

**Next step**: Add the numbers to the left of the reference base number 3
        2+2=4

Next Step: Multiply this addition in the last step by the reference base number 3.
        3x4=**12** (carry over is 1 here) --------------------------ii

Next Step: Multiply numbers to left of reference base number 3 and add Carry over 1
        2x2=4+1=**5**

Answer:    **529**

**Let us take one more example: 53x83**

First Step: Multiply 3by 3, we get: **9**----------------------------------------i

Next Step: Add the numbers to the left of 3; 5+8=13--------------------ii

Next step: Multiply 13 obtained in ii above by 3, the base reference number
        We get 39; keep **9** for use in the final answer, CO=3----iii

Next Step: Multiply units to the left of base reference number 3; we get
        5x8=40; add the CO=3 from above, we have **43**.

**Answer:**    **4399**
**Check:**    4+3+9+9= 7 (we ignore 9 and 9)
53=5+3=8
83=8+3=11
8x11=88; 8+8=16=7(16-9=7)
**Hence OK.**

**We must check our results no matter how obvious.**

**Let us move on.**


**IV--------------- Numbers ending in Digit 4**

Same procedure as for numbers ending in 2 and 3 applies.

But just to keep it in the picture, here is one example.

104x124: reference base number is 4. Numbers 10 and 12 are taken as units.

First Step: Multiply 4 by 4, we get 16: carry over is 1, keep **6** ---------
------i

Next Step: Add 10 and 12, multiply by reference base number 4 and add the carry over 1
          from first step, we get, we get 89; carry over 8 for the next step, keep **9**  ----ii

Next Step: multiply 10 by 12 and add the carry over 8 from last step, we get **128-----iii**
Answer: **12896**

Check:
104-----1+0+4=5
124-----1+2+4=7
 5x7=35; 3+5=8

12896=1+2+8+9+6=26-9-9=8 ( Remember anything above 9, subtract 9 and its multiples; in our case 18 is 9x2, so we let it go from 26.


There is absolutely no difference between the procedures for numbers ending in 2,3 and 4; same checks apply.

## V--------Numbers ending in 5

Here the attack takes on a much simpler approach.

You never ever will need a calculator if only you could bank on this number.

Example:

5X5=25 no big deal.
But 25x25=625, a little effort but not bad, we can do it easy in our head.
135x135=18225 here we need help.
Here is that help.

But like before, you cannot multiply any different numbers. You can only square quantities. It is a number midway between 0 and 10 and does not allow earlier applications to simplify computations beyond squaring the numbers ending in 5.

So let us take an example: 155x155: n=15, (n+1)=16: Base Reference is number 5.

First step: multiply 5 by 5, answer 25. Write it as it is always **25**--------i
Next step: Add 1 to 15, the number to the left of base reference 5; it is now 16---ii
Next Step: Multiply 16 by 15; so answer is **240**---------------------------iii
Next Step: Just write the answer as **24025** from iii and i above.

Check:

155=1+5+5=11
155=1+5+5=11
11x11=121=1+2+1=4------------------------------A

24025=2+4+0+2+5=13=13-9=4------------------B
A=B; hence the equation balances on both sides.

Bottom line: Write 25 with closed eyes and then add 1 to the number on the left of 5 in the problem; multiply using n(n+1) format, n being the first unit to the left of 5, n+1 being the second unit to the left of 5.

In our problem n= 15
  n +1=16

Always treat all numbers to the left of base reference as a unit only. Number 15 is to be looked as number 15 and not as 1 and 5.

---

## VI---------Digits ending in 6

Example 16x16: last digit here is 6 or our base reference is **6**
16------------a
X
16------------b
----
First Step: Multiply base reference digit 6 in 'a' by 6 in 'b'; we get 36; CO=3; keep 6 to
         the right for use in the final answer----------i
Next step: Add CO to the sum of units to the left of base reference number 6,
         (1+1)+ CO=2+3=5---------------------------------------------
--------ii
Next step: Using n (n+1) format, n=1 in 'a' and 'n+1'=1+1=2 in 'b'; units to the left of
         base reference number 6; 1x2=**2**-----iii

Answer: **256**

Another big example:

**186x186; Base Reference number is 6; n=18, n+1=19 and carry over is 'CO'**

First step: Multiply 6x6=36; CO=3; keep **6** for use in the answer--------------------i

Next Step: Add (18+18) plus CO=3; it becomes 39 CO=3, keep **9** for final use-----ii

Next step: Using again n(n+1) principle; **n=18, n+1=19**; it becomes 342 plus 3=345-iii

Answer= 34596

**Check**: 34596=3+4+5+9+6=27=9x3=0x3=0
186=15=6
186=15=6
6x6=36=3+6=9=0

---

## VII: Numbers ending in digit 7

Now we come to the number Seven. Seven is a complete number even in Heavens. God created the world in six days and rested on the seventh.

Let us take Numbers ending in 7.

Example: 177 X177

177------------------base reference number is 7 on the extreme right; n=17 on its left
  X
177------------------base reference number is 7 on the extreme right, n=17 on its left
------

First Step: Multiply 7 by 7, we get 49; keep 9 for later use and CO=4
Next Step: Multiply CO by n+1; Co=4 and n+1=17+1=18; we get 72; keep 2 for later use

in the answer and carry over 7 for next step.
Next Step: Multiply n by n+1; n=17 and n+1=18; 17x18=306; add CO from the last step
to it; CO =7, we now get 313.

Answer:    31329

Check: **31329**=3+1+3+2+9=18= multiple of 9; so it is **zero** as explained before.
**177**=1+7+7=15=15-9=6: anything above 9, deduct 9 or its multiples from it.
**177**=1+7+7=15=15-9=6
6X6=36=3+6=9=**0** as before.
Answer is correct as both equate to 9 or zero, as such equal in outputs.

## VIII-----Numbers ending in digit 8

Number 8 has a lot of significance to it. It is considered a very lucky number in China. A car with its license plate number as 888 was sold for a million dollars.

Let us deal with numbers ending in this digit 8.

Example 188x188

188-----------------8 is the reference base number and n=18 is the unit for our computation
  X
188-----------------8 is the reference base number and n=18 is the unit for our computation

First Step; Multiply 8 by 8; we get 64; CO=6 and keep **4** for later use in the answer
Next Step: Multiply CO=6 by n+1; n+1=19; we get=114;CO=11; keep **4** for use later
Next Step: Multiply n by n+1;n=18,n+1=19; we get 332; add CO=11 to it, we get

332+11=**343**.

Hence the answer is **343-4-4**.

Let us check

343444=3+4+3+4+4= 18=9x2=0---------------------A
188=1+8+8=17
188=1+8+8=17
17X17= 189=1+8+9=18=9x2=0----------------------B

A=B

Of course we have to learn to multiply small numbers in head or further break them down in the some way into such groups as above and deal on the same lines as we deal with lower values. Of course some of these would only work like those ending in numbers 5 or above when you can only square quantities, to that end, we may have to dig deeper for answers but still we can always check them without the calculators.

## IX----Numbers ending in digit 9

We are moving up. We are almost there. We will look into numbers that end in 9.

### Reference base number=9

199x199: our base reference number is 9. Numbers to its left are put in a unit. Whole 19 is one unit. We don't count it as 1 and 9. We designate this unit as n, its value being 19 in our problem below.

**199**------------ A  : **n=19; (n+1) = 19+1= 20**:  9 is the reference base number.
 X
**199**------------B
-----
39601

First step: Multiply 9 the base reference number by the other 9 in line B: CO=8, leave 1
for later use in the answer----------------------------------------------------------------i

Next Step: Multiply CO= 8 by n (n+1) of the unit; 8x20=160; CO=16, keep 0 for use in
the answer--------------------------------------------------------------
----------------------ii

Next Step: Now multiply n by (n+1); 19x20=380 and add to it CO=16 from above; so
We arrive at: 380+16=396--------------------------------------
----------------------iii

Answer: 39601

Check:  3+9+6+0+1=19= 19-18=1--------------------------------X

199=9+9+1=1(since 9 or its multiples are ignored)
199=9+9+1=1(Same Reason)
1x1=1-------------------------------------------------------------------Y

Hence **X=Y** or the answer is correct.

Happy computations!!!

**Digits ending in 0 are not that much of a problem.**

We would just let it go.
Utility of this little work is being able to check your calculations by simply adding the digits in the answer and comparing them to the answers obtained by using the procedures used in solving the problem as shown above under equations for X and Y.

And remember there is no substitute for a Human brain.

You can have all the calculators but should they fail, we must always have a back up and that is what this little book is all about..

Let us delve into a little Geometry now that has always been a part and parcel of mathematics..

Do we know we can convert an irregular Triangle into an Equilateral Triangle? It was first demonstrated by none other than Napoleon Bonaparte. If three Equilateral triangles are constructed off the sides of any triangle then the centers of circles which circumscribe each equilateral triangle are vertices of another equilateral triangle.

That is not the end of it. We can Trisect three angles of any triangle and see an equilateral triangle bloom in its midst.

**Napoleon's Theorem:**

Napoleon's theorem: If the triangles centered on $L$, $M$, and $N$ are equilateral, then so is the green triangle.

Napoleon's **theorem** states that if equilateral triangles are constructed on the sides of any triangle, either all outward or all inward, the lines connecting the centers of those equilateral triangles themselves form an equilateral triangle.

The triangle thus formed is called the inner or outer *Napoleon triangle*.

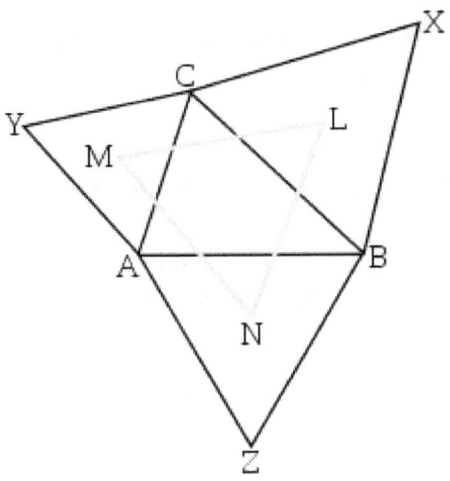

Napoleon Bonaparte's Theorem

Green Equilateral Triangle is born out of the three sides of the triangle under review. It can be done so many ways. Using compass we can circumscribe the three Equilateral Triangles and get the three points LMN.

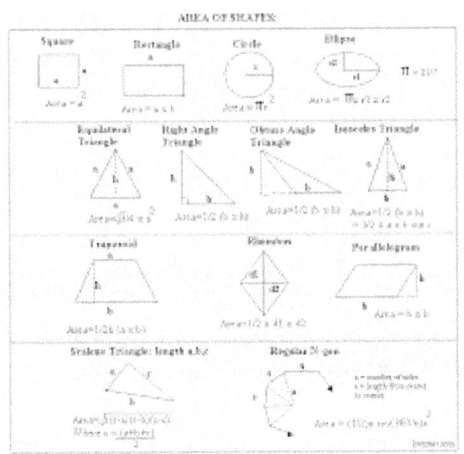

My concept of Geometry takes me to every Straight Line; every angle for Geometry has a wealth of Ideas. GEO-Me- try I guess means measurement of all that is on earth. Try me and see where I take you. From Plumb lines to its vast circuitous contours it seems to give us a large food for thought.

Leonardo da Vinci in his drawing of St. Jerome made it fit into a golden rectangle for he loved mathematics and its concepts.

In the words of Leonardo, " no inquiry can be called science unless it pursues its path through mathematical exposition and demonstration."

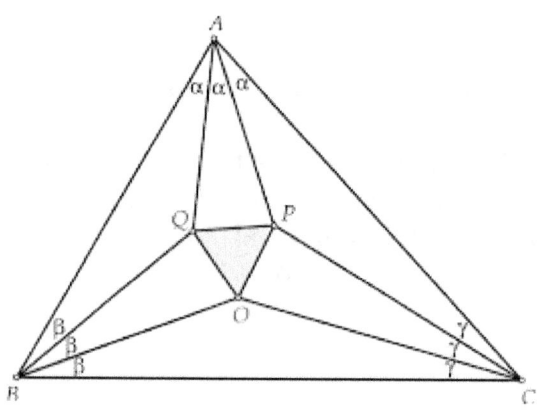

Morley's Tri-sector Theorem.

We can create an equilateral triangle from an irregular Triangle. We can undo any damage if we set our hearts to it. A wrong answer can be checked through simple means.

You have to go to the source rather than guess the problem from below. Author is on his way up to check Transformer details. **You have to scale the heights if you want to gain knowledge**

General:

Now that we are in the thick of things, we know we can multiply, divide, add, Subtract , let us go another way.

Here is a great mystery of mathematics if we can call it that.

Any Irregular triangle can be converted to an equilateral triangle.

There is nothing that is broken that cannot be put together. In fact I read Steve Hawkins's book (I do not remember the Title) where he says two pieces of broken china can be put together based on laws of conservation of Energy.

Ask me how.

Oral examinations for testing skills are now underway all around the globe for in oral examinations you are supposed to understand the concept beyond the theory and for that we have to have a little bit of mental storage of Data and its execution in good time. Time is of the essence today for every action and unless you have a little bit of understanding of the subject matter, we are only robots programmed for the tasks at hand.

In the world hierarchy of Mathematics we are lagging way behind; we are below even those third world countries that have no bread to eat, no shoes to wear; they cannot pull themselves up from their bootstraps as they say; we lag badly in basic and elementary mathematics. It is all because kids here are made to memorize numbers they have no knowledge of or in other words their implications in daily life. As an Engineer I have seen drawings showing a line of thirty feet smaller than a line of ten Feet in their layouts, a meter smaller than a Yard and so on and so forth on sealed drawings for crying out loud. But thank GOD, we have a natural bent of MIND that is HIS.

Not that far back on CNN, a one time mathematician, he had a degree in Mathematics was not able to recall the value of Pi ($\Omega$). It was so disturbing so disgusting to see it all, a man with a background of Mathematics ignorant of such a basic thing as value of Pi ($\Omega$). But that is the story of folks who only memorized all values without ever going around the circumference of the circles they drew.

In examinations for granting licensure today for various disciplines, there is so much deceit one is surprised things are still under control in most cases. There was an article in a magazine about a decade back on this topic; it pointed its fingers to major lapses in our testing and supervision; it said a few Groups offer review courses on various subjects and distribute marked pencils that have all answers to questions set in the Test Paper; these answers are etched or engraved around these pencils. If such practices are allowed in such high level testing, where should we go to find a level of sanity? Those not familiar with current methods of assessing competence may like to know how it all can come to pass. Bound volumes with complete paraphernalia are given to each student who takes their review courses. Gone are the days when we had to literally fill Reams of paper to answer a question, today all exams are computerized and answers are based on checking any of the five choices, Viz. A, B, C,.D or E. And they say if you do not know the answer or if the Pencil is not clear about the right answer, you may hit 'D'; you can be 25% sure it was the right answer. Believe it or not, it is true.

There are thousands of examples of cheating in the examinations but nothing exceeds this one; in testing for Textile Engineering, students are allowed to bring reference charts or reference tables to the examination hall; this is part of the process. Students literally bring solved examples and all other help material in full knowledge of the invigilators who may or may not be ignorant; they are drawn from the administrative staff of the College or the University. They have their doubts but they like to err on the side of the errors.

Back in 1983 in the Licensure examination for Professional Engineers, six of the Eight questions that were in the Test had been gone over in the review course with solved and ready to write solutions in the test; these were given in Bound volumes to all who took the review course given by none other than a University Professor prior to the date of examination. All questions were fully answered in their bound note books word for word to the last details;

like they say they were shovel ready projects. In most licensure examinations or tests for Professional Disciplines a select Committee consisting of college professors and general elders review the questions before they are finally set for the examinations to see if they fit the curriculum or the syllabus; now you know the rest of the story as Paul Harvey would say in his satires on his radio broadcasts. There was a hue and cry in the hall after the question papers were distributed for the examinations to begin but no action was taken despite a plethora of complaints to the Board by scores of people. My Friend who told me all about it was one taking the test in the same hall that year. He has committed suicide since though not for that reason or on those grounds; he was very angry and cursed the system in no uncertain terms. If such Licensures can be granted knowing there had been a let up, GOD knows where we would end up in lesser areas where perhaps there is not that much scrutiny.

Naval Academy is not immune from such inroads of cheating where naval Officers are commissioned to serve their country; one cadet refused to use any such help outside the one permitted and he did not graduate. He was on CBS sixty minutest TV show telling it all to the surprise of his peers.

One may say where it all comes in this kind of book. Elementary, Dr. Watson as Sherlock Home would say, it tells on our basic integrity and understanding of education, its very purpose. Mathematics is an absolute subject where two and two always would make four, it cannot produce fakes since fakes become teachers at some at point in Life and they impart a wrong purpose for learning the subject matter. If Education only makes robots of kids , we do not need live Robots, Electronic Robots are doing a much better job, they do not complain and are good to the last digit. Mathematics teaches integrity, the very lifeline of Life.

It is the process of learning that gives us a kick. As you add or Subtract, it is not just some abstract numbers we are dealing with, we can feel something in our bones and when we get it right and check it too, it gives an 'Einsteinium' feeling we are not that far behind him either.

Gauss once wrote "It is not knowledge, but the act of learning which grants the greatest enjoyment. When I have clarified and exhausted a subject, then I turn away from it, in order to go into darkness again".

**We must understand these floating numbers and symbols even if we do not use them**

We need to know the significance of each symbol above. Our modern day calculators are so well programmed that unless you are a dud, you can pass off as a genius. It is like a singer singing a song by only syncing his or her lips. It happened here where milli Vanilli got the award for a great song when they barely could sing otherwise. They were caught, that was music; imagine someone faking a critical area and be a danger to our very way of life and safety like in a control room of a nuclear plant or worst still in a repository of Long Range Missiles or any other such location.

In money exchanges across the Globe, the so called traders of currencies" know very little or fake it so; they disburse and count the conversions all to their advantage; these 'sickos' of Society deceive old people all day long. They should be put away for Life; we do not have to give them any slack. We should know our numbers.

A thief thinks he has gained a lot as he steals. Little does he know in the end it all balances out!! Only pain in all these transactions is the victim loses not only his money but his peace of mind; nobody wants to suffer the pain of having been cheated in broad day light. Again, learn to count way before others learn to cheat. Oral Mathematics is the key.

GOD is our witness. HE does restore all we lose.

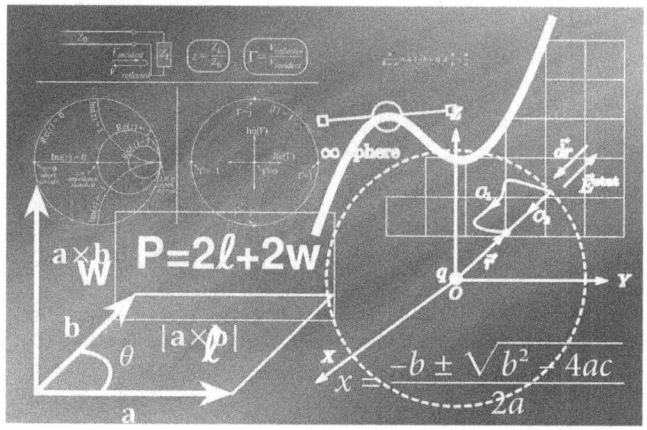

Plug in your numbers and know the outcome.

Six Great Mathematicians:

Pierre de Format  
Archimedes

Sir Isaac Newton

Bernhard Riemann          J Lagrange          Carl Gauss

Each one of them started from square one and went on to create a legacy that to this day lives. The road to success starts from step one; Journey of a thousands mile starts with the first step though at times it may not be necessary.

There is a book Acres of Diamonds that says Diamonds are all there in your own acres. You do not have to venture out looking for them. But base of all that is you must be true to yourself, to your GOD who created you and last but not least never pose as somebody you are not.

We often display two sides of our personalities each different from the other; we are the same outside as we are inside; in the process we lose our integrity. A kid who knew Two plus Two always is four began to count it wrong as he got mixed up in life.

Mathematics is **the** litmus test of your mind. As long as you can count it right, you are all clear and all is quiet on the western front. You may miss on the facts of history, it can be condoned but once you miss on a number you always knew, it is deemed serious.

No invention is complete without any analysis. A Missile Test is good only if it reaches its intended distance; if it misses a few miles, it is failure though it may still do the damage. What is right is right, what is wrong is wrong. It is like those Logic gates where either it is '0' or it is '1' defining if the switch is open or closed, no two ways about it.

Mathematics is the language of Numbers. Like they say, a picture is worth a thousand words.

The Six stalwarts of the field of Mathematics posted above mean so much to our present day world of Mathematics. There are so many of them and were I to list them all, this computer will bust or crash; their contributions are never away from any grateful mind. They developed all the basics we bank upon today when there was very little help from anywhere; they checked and rechecked their work solely on their own using their fertile brains and their logic.

Sir Isaac Newton

Newton is so famous for his calculus, optics, and laws of gravitation and motion, it is easy to overlook that he was also one of the very greatest geometers. Despite the power of Descartes' analytic geometry, Newton's achievements with synthetic geometry were surpassing. An anecdote often cited to demonstrate his brilliance is the problem of the '*brachistochrone*', which had baffled the best mathematicians in Europe. It came to Newton's attention late in life. He solved it in a few hours and published the answer anonymously. On seeing the solution Jacob Bernoulli immediately exclaimed **"I recognize the lion by his footprint." Newton was a lion by all standards.**


Newton ranks #2 on Michael Hart's famous list of the Most Influential Persons in History. Whatever the criteria, Newton would certainly rank first or second on any list of physicists, or scientists in general. One reason he may have ranked him at #1 is a comment by Gottfried Leibniz himself: "Taking mathematics from the beginning of the world to the time when Newton lived, what he has done is too great to ignore. "

Bill Gates in the modern era does deserve mention as he introduced us to our laptops and compressed the whole world to a few inches.

Carl Gauss

Carl Friedrich Gauss, the "Prince of Mathematics," exhibited his calculative powers when he corrected his father's arithmetic before the age of **three.** His revolutionary nature was demonstrated at age **twelve,** when he began questioning the axioms of Euclid. His genius was confirmed at the age of **nineteen** when he proved that the regular n-gon was constructible if and only if it is the product of distinct prime Fermat numbers. Also at age 19, he proved Fermat's conjecture that every number is the sum of three triangle numbers. At age **24** he published *Disquisitiones Arithmeticae*, probably the greatest book of pure mathematics ever.

Gauss once wrote "It is not knowledge, but the act of learning which grants the greatest enjoyment. When I have clarified and exhausted a subject, then I turn away from it, in order to go into darkness again ...".

Archimedes

Archimedes is universally acknowledged to be the greatest of ancient mathematicians. He studied at Euclid's school. His achievements are particularly impressive given the lack of good mathematical notation in his day. No calculators or any other help.

Archimedes was an astronomer. He was one of the greatest mechanists ever, discovering **Archimedes' Principle of Hydrostatics**. The story that he determined the proportion of gold and silver in a wreath made for Hieron by weighing it in water is probably true, but it was **Galileo** who pointed out that a test based on measuring water displacement, as had been assumed to be Archimedes' "Eureka!" method would be extremely imprecise.

He never lost sight of observation, Experiment and calculation. This same principle holds in our Healings today. We face and combat sickness, death or disease much the same way as young interns learn from their peers; they observe, experiment, calculate and learn to draw their own conclusions. Archimedes was a devout observer and experimenter.

Galileo

Galileo later defended his views which appeared to attack Pope John Urban VIII thus alienated him and the Jesuits who had both supported Galileo up until this point. He was tried by the Inquisition, found "vehemently suspect of heresy", and forced to recant. He spent the rest of his life under house arrest. While under house arrest, he wrote "Two new Sciences" in which he summarized work he had done some forty years earlier on the two sciences now called **kinematics** and **strength of materials**.

Lagrange

Joseph-Louis Lagrange was a brilliant man who advanced to become a teen-age Professor shortly after first studying mathematics. He excelled in all fields of analysis and number theory; he made key contributions to the theories of **determinants,** continued fractions, and many other fields. He developed partial differential equations far beyond those of D. Bernoulli and d'Alembert, developed the calculus of variations far beyond that of the Bernoulli's, discovered the Laplace transform before Laplace did, and developed terminology and notation (e.g. the use of $f'(x)$ and $f''(x)$ for a function's 1st and 2nd derivatives).

Unlike Newton, who used calculus to derive his results but then worked backwards to create geometric proofs for publication, Lagrange relied only on analysis. "No diagrams will be found in this work" he wrote in the preface to his masterpiece work.

Lagrange once wrote "As long as algebra and geometry have been separated, their progress has been slow and their uses limited; but when these two sciences have been united, they have lent each mutual force, and have marched together towards perfection."

You think they had these big computers, no, they used their brains.

Let us lighten up and see how cool mathematics is. Everything in Mathematics goes up to 9 and then it repeats. We make every digit add up to 9 after Bob spits out his Numbers. .

Pascal

Blaise Pascal was France's most celebrated mathematician and physicist and religious philosopher. He was a child prodigy who was educated by his father. He worked on conic sections and projective geometry and he laid the foundations for the theory of probability. In 1642, at the age of 18, Pascal invented and build the first digital calculator as a means of helping his father perform tedious tax accounting. Pascal's father was the tax collector for the township of Rouen.

All these great mathematicians developed their machines from basic numbers and used then only after they had validated their inventions through checks and cross checks. We tend to go backwards; we use the machines as our first line of defense and then settle down to plug in our inputs.

Here is a real story. How we do not trust our own brains. Even our computers have picked on it as they say, 'Did you mean' "whatever". My colleague was staying in a Posh Hotel someplace on this planet; he was booked for three days. On second night Front Desk sent security to ask him to leave in the middle of the night. He was stunned. They went down together; the computer showed that he was to leave on the second night. The Clerk would not budge saying the computer cannot be wrong; the Guest had a receipt. Where do you go in such cases where Computer is truer than the man and his paper..

That is why Every Software needs to be validated before it is commissioned.

Because Paper writing material was at a premium during the middle ages, counting and communicating results by finger signs were often used. These were fairly well understood.

We must always keep an eye on Roman symbols; they become tricky after a while.
 Hindu and Mayan symbols did not change much. We can always look them up but we must know them by heart and not begin to surf in the heat of the moment.

Old Alien Stories or old TV Shows write their Serials in Roman Symbols; even our Super Bowl Serial Number is written in Roman Symbols (L1V). They say that in the End Rome is coming back as super Power, I do not know but we use lot of their inscriptions nd disciplines from Senate to our Latin masses.

## *Roman numerals*

| Number | Roman numeral |
|---|---|
| 0 | not defined |
| 1 | I |
| 2 | II |
| 3 | III |
| 4 | IV |
| 5 | V |
| 6 | VI |

| Number | Roman numeral |
|---|---|
| 7 | VII |
| 8 | VIII |
| 9 | IX |
| 10 | X |
| 11 | XI |
| 12 | XII |
| 13 | XIII |
| 14 | XIV |
| 15 | XV |
| 16 | XVI |
| 17 | XVII |
| 18 | XVIII (Higher symbol X); So number is 18). |
| 19 | XIX |
| 20 | XX |
| 30 | XXX |
| 40 | XL |
| 50 | L |
| 60 | LX |
| 70 | LXX |
| 80 | LXXX (A higher symbol on left increases value.) |
| 90 | XC (A lower symbol on left decreases its value by that amount) |
| 100 | C |

Little Sum:

Let us say one is floating some numbers saying we can make millions if we go along with him.

Let me regale you with a little trick Problem and Solution. Supposing Bob gives you a Five Digit Number. You can tell him you can post an answer (i) after two more of his statements of Five Digits. Our Answer will remain the same no matter what he might quote in his two more inputs.

Let us take one example. You are sitting in a coffee House waiting for your order. You can wait it out if you are with Bob and play it away.

It is worth its time.

```
              1  5  5  5  5 --------------------Bob -----i
              2  0  5  9  5--------------------Bob------ii
              7  9  4  0  4------------------- Your input----ii
              3  9  8  7  5-------------------- Bob-----------iii
              6  0  1  2  4------------------ Your Input----iii
        -----------------------------------------------------------
                    215553-------(i) your answer after Bob's
```
first call.

Let us try one more time.

```
              5  6  7  8  3       Henry--i
              4  3  6  7  8       Henry--ii
              5  6  3  2  1       Mike---ii
              1  3  8  6  4       Henry—iii
              8  6  1  3  5       Mike—iii
        -----------------------------------------------------------
                    2 5 6 7 8 1              Mike---i
```
(Answer)

Mike writes the **answer** 256781 as above on Henry's First Number of Five Digits. Henry then inputs his second Number. We make each digit add up to 9. Henry inputs his Third input. We do the same again; add it all up to make them 9.

What we did is we subtracted 2 from the last digit; so 3_2=1. Write it out 1 like we did above.

Write the rest of the digits as these were. So we wrote 5,6,7,8 as they were.

Then put the number 2 after writing those Four above digits.

Thing to remember is number '9' is our Chief for all inputs.

We see that TV show where kids come swith answers to even most varied and complex numbers as the Host asks how much would 16667778899x189768543879 would be.

Everything in our world is well laid out; we only have to discover it.

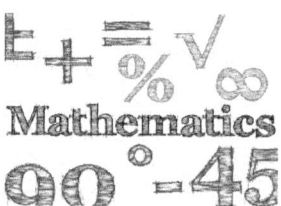

Daily use symbols

It may look primitive but it does come handy at times. In Fact there is a chart. it goes all the way into thousands.

Even to this day, villagers resort to finger counting. It is the surest way to count your way out of trouble. They close one as they open the other and so on and so forth. Our whole destiny is written on our hands.

**Prime Numbers**

|  |  |  |  |  |  |  |  |  |  |
|---|---|---|---|---|---|---|---|---|---|
| 1 | 2 | 3 | 4 | 5 | 6 | 7 | 8 | 9 | 10 |
| 11 | 12 | 13 | 14 | 15 | 16 | 17 | 18 | 19 | 20 |
| 21 | 22 | 23 | 24 | 25 | 26 | 27 | 28 | 29 | 30 |
| 31 | 32 | 33 | 34 | 35 | 36 | 37 | 38 | 39 | 40 |
| 41 | 42 | 43 | 44 | 45 | 46 | 47 | 48 | 49 | 50 |
| 51 | 52 | 53 | 54 | 55 | 56 | 57 | 58 | 59 | 60 |
| 61 | 62 | 63 | 64 | 65 | 66 | 67 | 68 | 69 | 70 |
| 71 | 72 | 73 | 74 | 75 | 76 | 77 | 78 | 79 | 80 |
| 81 | 82 | 83 | 84 | 85 | 86 | 87 | 88 | 89 | 90 |
| 91 | 92 | 93 | 94 | 95 | 96 | 97 | 98 | 99 | 100 |

Prime Numbers.

Let me sum up all thus far. We covered lot of ground. We went from multiplying simple numbers ending in 1 thru 10 to showing a few crooked geometry figures to prove there is nothing that cannot be fixed.

If you drew an irregular triangle by mistake and you have no eraser, you can convert it into an equilateral triangle without flexing your muscles. You need not even have a compass, a straight edge would do.

A little exercise to break the monotony by showing that number 9 is the trump card. It is the peak of numbers in daily solutions.

Lest I forget do not ever lose sight of Binary numbers, they are the lifeline of all logic. You are either wrong or you are right.

A binary number is generally much longer than its corresponding decimal number; for example, 256,058 has the binary representation 111 11010 00001 11010. The reason for the greater length of the binary number is that a binary digit distinguishes between only two possibilities, 0 or 1, whereas a decimal digit distinguishes among 10 possibilities; in other words, a binary digit carries less information than a decimal digit. Because of this, its name has been shortened to bit; a bit of information is thus transmitted whenever one of two alternatives realized in the machine. It is of course much easier to construct a machine to distinguish between two possibilities than among 10, and this is another advantage for the base 2; but a more important point is that bits serve simultaneously to carry numerical information and the logic problem. **That is, the dichotomy of yes and no, and of true and false, are preserved in the machine in the same way as 1 and 0, so in the end everything reduces to a sequence of those two characters.**

And let us end it all on a happy note. Let the wicked not use compromised weights and cheat the poor and defenseless and weigh less than the poor pays for it; let the killer not use his sword instead turn it as a plowshare and sow wheat for the poor. Let the education be more for dusting off the dirt rather than piling on the ill gotten gains in our safe vaults.

Let us remove the dross from around us and come out as clean and repentant humans as we come out of the ovens of our misdeeds.

We have come a long way from the days of Abraham, Buddha and JESUS. We may not use Abacus today but we cannot forget what it taught us. We have jumped from Abacus to Calculators in little over two milleniums; let us never forget that it is little things that teach us how to live.

And finally you must never have to look into your notebooks to remember Telephone numbers of folks you love. Remember the area code, if you forget that, God Save the Queen. Learn to group like numbers in your head. Like for: 000- 380-5800, the group is in 38, 58 etc. For 000-49-7059, the group is 49, 59 etc. You can compress the range into a number or two.

This little effort is a step in that direction.

We only talked of Multiplying.

We can **add/Subtract** in the same way with little change.

Take any Addition.

Say 15+15=30
Check: 1+5=6-----------------------i
1+5=6 ------------------------------ ii
And add these two= 6+6=12-----iii.
12=1+2=3   LHS---------------------iv
RHS: 30=3+0=3; So 3=3-----------v.

LHS=RHS

---

We can take a bigger number.

{1456789}= sum of all digits=40=4+0=4 : (add all digits 1+4+5+6+7+8+9=40)
     +                                                  =
4+6=**10**=9+1=1(we count 9 as 0)—(i)
{3456789}= Sum of all digits=42=4+2=6 : (3+4+5+6+7+8+9=42);

Total= 4913578  = sum of all digits=37=3+7=**10**=9+1=1(we count 9 as 0) ------------(ii)

10=10
1=1
Left Hand Side= Right Hand Side

The central theme in all these iterations is stick to number 9, make sure your Left hand side comes to the same total as your Right Hand Side.

**Famous Mathematicians of the 20th Century**

- Katherine Johnson. 26 August 1918, American. ...

- Alan Turing. 23 June 1912, British. ...
- John Forbes Nash Jr. 13 June 1928, American. ...
- John von Neumann. 28 December 1903, Hungarian, American. ...
- Shakuntala Devi. 04 November 1929, Indian. ...
- Tom Lehrer. 09 April 1928, American. ...
- Paul Dirac. 08 August 1902, British. ...
- Kurt Gödel.
- There are many more but these ones have made a huge difference in the Life and Struggles of humanity. Missing from list is Laplace; a French Scholar. No he was no less of a mathematician than ones in the list above; I just forgot. Who can forget Laplace Transform?

Let me cap it all with my Indian genius whose work took the then British Scholars by storm.

Srinivasa **Ramanujan.**

Srinivasa **Ramanujan**, Erode, **India**
**Indian mathematician**: Contributions to theory of numbers include pioneering discoveries of the properties of the partition function.
Born: December 22, 1887, Erode
Education: University of Madras, Trinity College

He worked out the Riemann series, the elliptic integrals, hyper geometric series, the functional equations of the zeta function and his own theory of divergent series.

I think I read he was a genius whose talent was never brought out in those days of British rule and social distancing (race relations) in the world. They say they found scrolls of papers on calculus and other areas of his expertise in the Cambridge Library as they were dusting its shelves. They were bundled over and in bad shape.

Hardy is somewhere in this Picture.

In 1916 **Ramanujan** got his BA from **Cambridge** and his research went from strength to strength. He published one excellent paper after another, with a great deal of Hardy's helping the proofs and presentation. They also collaborated on several great projects, and published wonderful joint papers.

Hardy was not the first mathematician to whom Ramanujan had sent his results; however the first two to whom he had written judged him to be a crank. But Hardy was not only an outstanding mathematician; he was also a wonderful teacher, eager to **nurture talent.**
Anyone with a honest bent of mind can scale heights he or she could never have imagined. It is like Joel Osteen, a TV Preacher at Lakewood Church in Houston, Texas, saying GOD will take you to places which you never imagined. Mathematics is one of perhaps a few of the subjects that is not only useful and helpful in day to day Life; it is the very base of our existence.

John Adams was an American statesman, attorney, diplomat, writer, and Founding Father who served as the second president of the United States, from 1797 to 1801, proved Pythagoras Theorem sitting in his office by drawing squares on each of the three sides of a right angle triangle and counting them all ; proving that the Square on the Hypotenuse is equal to the sum of the squares on the other two sides.

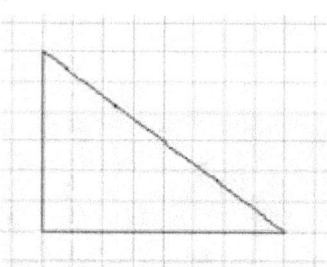

**For a 3, 4, 5 Right angle triangle, the hypotenuse would have 25 squares and the other two would have 9+16 =25 Squares.**

Not to belittle or ignore other areas of Life, it will not be out of the way or even out of order to say that if you can master basics of Mathematics, you can master anything for in its absolute dictates, there lingers no dioubt.

www.ingramcontent.com/pod-product-compliance
Lightning Source LLC
Chambersburg PA
CBHW050301220526
45465CB00002B/775